# Our Solar System

## Young Explorer Series: Book 1

ISBN 978-1-7352616-0-7

DLK Publishing

825 N. Gould St.

Owosso, MI   48867

www.dianakanan.com

# Dedication

For my little astronomer, Oliver. My wish is that you always shoot "past" the stars!

With love, Grandma

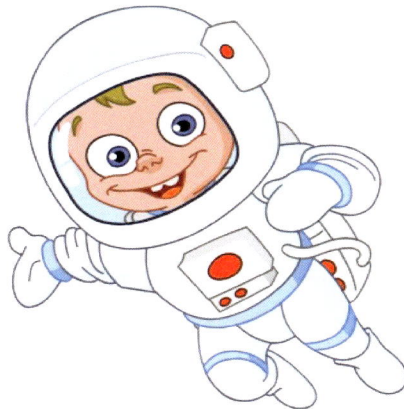

# Our Solar System

Our solar system is made up of the Sun and everything that orbits around it, including the planets- Mercury, Venus, Earth, Mars, Jupiter, Saturn, Uranus, and Neptune.

Our solar system also contains dwarf planets, such as Pluto, dozens of moons and millions of asteroids, comets, and meteoroids.

# SUN

The Sun is a huge, spinning, glowing ball of hot gas that sits in the center of our solar system. The Sun's gravitational pull keeps the planets in an orderly plane as they rotate around it.

# MERCURY

The planet closest to the sun is Mercury. It is also the smallest planet.

Although Mercury's surface is similar to Earth, the temperatures are quite different. Because of its thin atmosphere its temperature can be as high as 800 degrees and as low as -290 degrees. Under those extreme conditions, it is unlikely that life could develop on Mercury.

Mercury has no satellite (moon).

FUN FACT: A year on Mercury is only 88 days.

# VENUS

Even though it is not the closest to the Sun, the thick atmosphere of Venus traps in heat making it the hottest of all the planets.

It is so hot on Venus, metals like lead would be puddles of melted liquid. The extreme temperatures make it an unlikely place for life to exist as we know it.

Venus does not have a moon.

**FUN FACT:** Because Venus is close in size to Earth it is sometimes called our sister planet.

# EARTH

The Earth is the third planet from the Sun. It is the planet we call home and the only planet known to support life.

The Earth is mostly iron, oxygen and sillicon. It is 71% water and 29% land. Water is an essential ingredient for life to exist on our planet.

The Earth has one moon.

**FUN FACT: The Earth is 4.54 billion years old.**

# MARS

The fourth planet from the Sun is Mars. Sometimes known as the "red" planet due to the effects of iron oxide on its surface.

The temperatures range between -195 degrees and -80 degrees.

Recent visits to Mars have discovered a sub-glacial lake below the surface of the southern polar cap.

Mars has two moons.

**FUN FACT: Mars is the most explored planet in our solar system.**

# JUPITER

Jupiter, the fifth planet from the Sun, is the oldest and the largest of all the planets.

There is no surface on Jupiter as it is made up of gas and liquids.

Temperatures have been estimated in the center of Jupiter to be hotter than the surface of the Sun.

Jupiter is the fourth brightest object in the solar system-only the Sun, Moon and Venus are brighter. It is one of five planets visible to the naked eye from Earth.

FUN FACTS: A day on Jupiter is 9 hours and 55 minutes. Jupiter has 79 moons.

# SATURN

Saturn is the sixth planet from the Sun. It is the most recognized planet because of it's destinctive rings.

The rings are made up of billions of particles that range in size from tiny dust grains to objects as large as mountains.

Saturn has 82 moons, more than any other planet in the solar system.

**FUN FACT:** Saturn's density is lighter than water, meaning it would float if put in water!

# URANUS

Uranus is the seventh planet from the Sun. It is made up of water, methane, and amonia with an atmosphere of mostly hydrogen, helium, and methane gases.

Methane is what makes Uranus appear blue.

The temperature on Uranus is estimated to be -371 degrees. It is often described as an "ice planet."

Uranus has 27 moons.

**FUN FACT:** Uranus and Venus are the only two planets that spin east to west.

# NEPTUNE

The eighth and furthest planet from the Sun is Neptune.

Neptune is another "ice planet" similar to Uranus, it is made up gases.

Neptune has winds that whip clouds across the planet at speeds of more than 1200 mph-close to the speed of a U.S. Navy jet fighter.

Temperatures average -375 F.

Neptune has 14 moons.

Fun Fact: A spacecraft traveling from Earth to Neptune would take ten years.

Scientists are learning more and more every day as they develop more power telescopes to explore the atmosphere.

They also learn by man-made machines called satellites that are sent into space to take pictures, not only of our Earth, but other planets, the sun and other objects.

Satellites fly high in the sky, so they can see large areas of Earth at one time. Satellites have a clear view of space. That's because they fly above the Earth's clouds and air.

The first human to journey into outerspace was a Russian cosmonaut, Yuri Gagarin. He made one orbit around our planet in 1961.

Nine years later, July 20, 1969, the United States made history by landing on the moon. During that expedition, astronaut Neil Armstrong became the first human to step on the moon.

He, and astronaut Buzz Aldrin, walked around for three hours as they picked up bits of moon dirt and rocks for scientists to study back on Earth.

# The Earth's Moon

Our moon is the brightest and largest object to us in the night sky. It is the closest heavenly body to Earth and can easily be seen with the naked eye. For generations it has been a constant reminder of the mystery of what lies beyond.

The moon's surface is like a desert with plains, mountains, and valleys. It also has many craters created from space rocks that hit the surface at a high speed. There is no air to breathe on the Moon.

The last time man visited the moon was in 1972.

# Word From the Author

One of my favorite memories as a child was laying in the grass, on a hot summer night, starring up into the stars. To this very day, I still experience that same feeling of "awe" when I imagine what could lie beyond what we can see.

# For more information on the

# Young Explorer Series
## visit

## www.dianakanan.com

www.ingramcontent.com/pod-product-compliance
Lightning Source LLC
Chambersburg PA
CBRC101142030426
42336CB00007B/70